NATURE CLOSE-UP

TEXT BY ELAINE PASCOE

PHOTOGRAPHS BY DWIGHT KUHN

BLACKBIRCH PRESS, INC.

WOODBRIDGE, CONNECTICUT

Published by Blackbirch Press, Inc.
260 Amity Road
Woodbridge, CT 06525

To my nephew Brian K. –D.K.	To P.B. –E.P.

First Edition

Email: staff@blackbirch.com
Web site: www.blackbirch.com

Printed in the United States

10 9 8 7 6 5 4 3 2 1

Photo Credits: All photographs ©Dwight Kuhn, except pages 25, 34–39: ©David Kuhn/Dwight Kuhn Photography.

front cover: adult horse fly
back cover: black fly eggs, black fly larva, black fly pupae, adult black fly emerging from its pupal skin

Library of Congress Cataloging-in-Publication Data

Pascoe, Elaine.
Flies / by Elaine Pascoe; photographs by Dwight Kuhn
p. cm. — (Nature close-up)
Includes bibliographical references (p. 47).
Summary: Describes the characteristics, habits, life cycle, and appearance of the many species of flies. Includes experiments.
ISBN 1-56711-149-1
1. Flies—Juvenile literature. 2. Flies—Experiments—Juvenile literature. [1. Flies. 2. Flies—Experiments. 3. Experiments.] I. Kuhn, Dwight, ill. II. Title.
QL533.2.P37 2000 99-053769
595.77—dc21 CIP
AC

Note on metric conversions: The metric conversions given in Chapters 2 and 3 of this book are not always exact equivalents of U.S. measures. Instead, they provide a workable quantity for each experiment in metric units. The abbreviations used are:

cm	centimeter	**kg**	kilogram
m	meter	**l**	liter
g	gram	**cc**	cubic centimeter

CONTENTS

1

Flies Everywhere

What do you do when a fly buzzes by? You probably shoo it away. Next time, take a minute to watch it. Even though flies can be pests, these insects deserve a closer look.

Flies are true aerial acrobats—they can perform amazing maneuvers in the air. They're pretty quick on their feet, too. Flies can even walk up walls and across ceilings! And flies are as adaptable as they are acrobatic. They are found in almost every part of the world in every kind of habitat—deserts, swamps, shorelines, woodlands, suburbs, and cities.

Flies are highly adaptable insects with amazing flying abilities.

THE FLY FAMILY

When you think of a fly, you probably picture a housefly. Houseflies are very common and are found wherever people live. But there are more than 85,000 different kinds, or species, of flies. Most flies prefer warm weather and can't survive in freezing temperatures. But some have adapted to harsh conditions. The larvae (immature forms) of certain flies even make their homes in hot sulfur springs and pools of crude oil!

Besides houseflies, the most common members of the fly family include blowflies and tiny fruit flies. Mosquitoes are one of the many kinds of biting flies. Other biters include black flies (also called buffalo gnats), horse flies, and deer flies. People avoid these flies because their bites are painful, and because they can spread diseases.

Horse flies are one of the many kinds of biting flies.

Left: **The head of a fly contains its mouth and sense organs, which include large eyes.** *Below:* **Like all insects, a fly's body has three sections—the head, thorax, and abdomen.**

With so many members, the fly family includes a great number of body types. Flies range in size from "no-see-ums"—tiny midges less than 1/16 inch (1.6 mm) long—to huge robber flies, which can be more than 3 inches (76 mm) long. Some flies are long and thin, while others are short and plump.

Like all insects, a fly has six legs and a hard outer covering called an exoskeleton. Its body has three segments. The head holds the fly's mouth and sense organs. The center segment, or thorax, is mainly made of muscle. The wings and legs attach to the thorax. The abdomen contains the fly's digestive and reproductive organs. Flies breathe through small holes called spiracles, located on the abdomen and thorax.

Above: **Flies have only one pair of wings, unlike other insects, which have two pairs.**
Below: **The sticky pads on a fly's legs help it to walk on slippery surfaces like this glass thermometer.**

Wings are the feature that sets flies apart from other flying insects. Most flying insects have two pairs of wings. A fly, however, has only one pair. The wings, which are generally transparent, don't look so remarkable. But they're very powerful. A housefly's wings beat about 200 times a second. The wings of some midges beat five times faster! In fact, midges hold the record in the insect world for the fastest wing beats.

In place of a second set of wings, a fly has a pair of stalk-like organs called halteres, located just behind the wings. They help the fly balance during flight. The halteres detect every roll and pitch as the fly zooms along. With their help, the fly always knows which way is up and can correct its flight path.

A fly's six legs end in claws that help it cling to rough surfaces. At the base of these claws, many flies have two or three small, sticky pads. The pads help flies walk on smooth surfaces, such as glass. Flies also have taste receptors on their feet. They can actually tell if something is good to eat by walking on it!

WHEN IS A FLY NOT A FLY?

Every insect with "fly" in its name isn't a member of the true fly order, which scientists call *Diptera*. For instance, fireflies, which flash and glow on summer nights, are actually a kind of beetle. So is the American Spanish fly. Sawflies are also not true flies; they belong to a different insect family. Caddisflies, damselflies, dragonflies, mayflies, and scorpionflies are other so-called "flies" that belong to separate insect families.

Dragonfly

ALL EYES

The most obvious feature of a fly's head is its eyes. Most flies have two kinds of eyes: compound and simple. The two compound eyes are so big that they take up most of a fly's head. Each eye is made up of many lenses, or facets. In some species, there are as many as 4,000 facets in each eye. Simple eyes have only one facet each. Many flies have three small simple eyes grouped at the top of their heads.

Huge compound eyes dominate a fly's head. Certain flies have eyes that contain about 4,000 separate facets, or lenses.

Long, feathery antennae branch out from the head of this male mosquito.

Despite its five eyes and thousands of lenses, an average fly probably doesn't have very good vision. Simple eyes mainly pick up changes in light levels, and compound eyes mainly detect motion. Flies cannot focus clearly on one object. The many lenses of compound eyes, however, allow a fly to see an object from many different angles and to react to movement quickly.

Female mosquitoes have long, slender antennae that serve as organs of touch and smell.

A fly's two antennae, or "feelers," are organs of touch and smell. They're located between the compound eyes. The antennae are sensitive enough to detect the movement of air currents, so the fly senses an approaching threat (like a fly swatter) before it's too close. Houseflies have a pair of short, thick antennae. Female mosquitoes have long, thin, feelers, while male mosquitoes have feathery ones. The male's antennae help him find a mate because they're sensitive to sounds created by the beating wings of females.

Below the eyes and antennae are the fly's mouthparts. Adult flies can eat only liquids, which they suck up through a tube-like structure called a proboscis. Many flies, including houseflies, have soft pads at the tip of the proboscis. These pads act like little sponges to draw liquids into the proboscis. Biting flies do not actually bite—they stab their victims with their sharp mouthparts.

Right: **A fly's mouthparts are located below the eyes. In between the eyes are short, thick antennae.**
Below: **A housefly's proboscis is used to suck up liquid food.**

Black fly eggs lie under water in a stream.

FROM EGG TO ADULT

A fly goes through four stages during its life: egg, larva, pupa, and adult. At each stage, the insect looks very different because its body changes. This process is called complete metamorphosis. A fly's life is short—some midges live only a few hours, and adult houseflies typically live just a few weeks. But there may be many generations of flies in one year.

Adult female flies lay from one to several hundred eggs at a time, depending on the species. The eggs are usually yellow or white, and look like seeds or grains of rice. Some flies just drop their eggs on the ground. Others push them into decaying material, such as rotting wood or manure. Mosquitoes and some other flies lay their eggs in water. Some mosquitoes even lay rafts of eggs, held together with a sticky substance.

In warm weather, housefly eggs hatch in as little as eight hours. But the length of time to hatching varies, depending on the fly species as well as the weather. Eggs laid in fall may not hatch until the following spring. Then the tiny worm-like larvae wriggle out to eat and grow.

Mosquito larvae live in stagnant (non-moving) ponds and pools. Black fly larvae live in running streams. The larvae of houseflies and blowflies live in spoiled food and other rotting material. They are sometimes called maggots. Some fly larvae are parasites—they live on or in plants or animals. For example, bot fly larvae burrow under the skin of animals and feed off their hosts.

A black fly larva, underwater

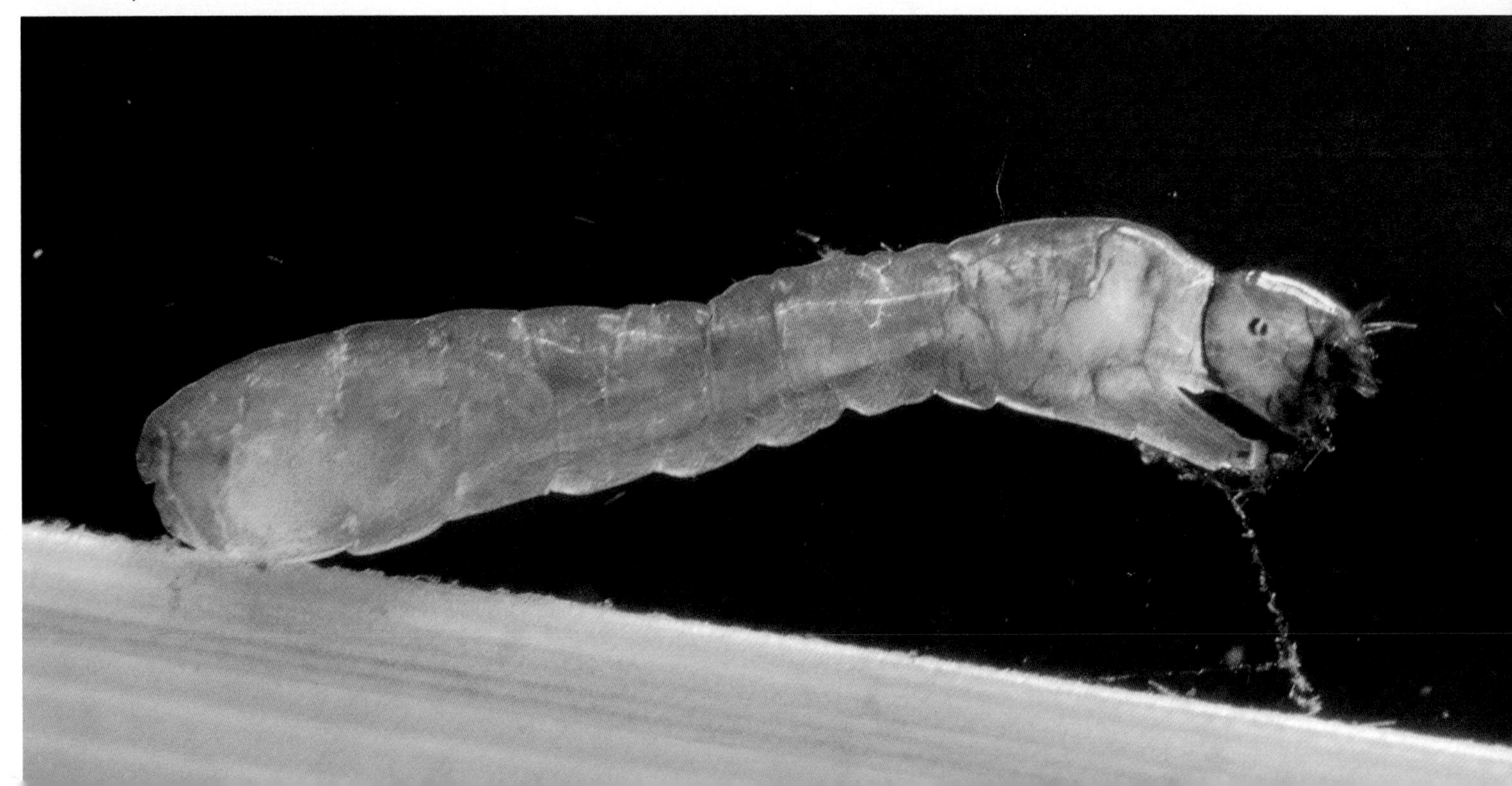

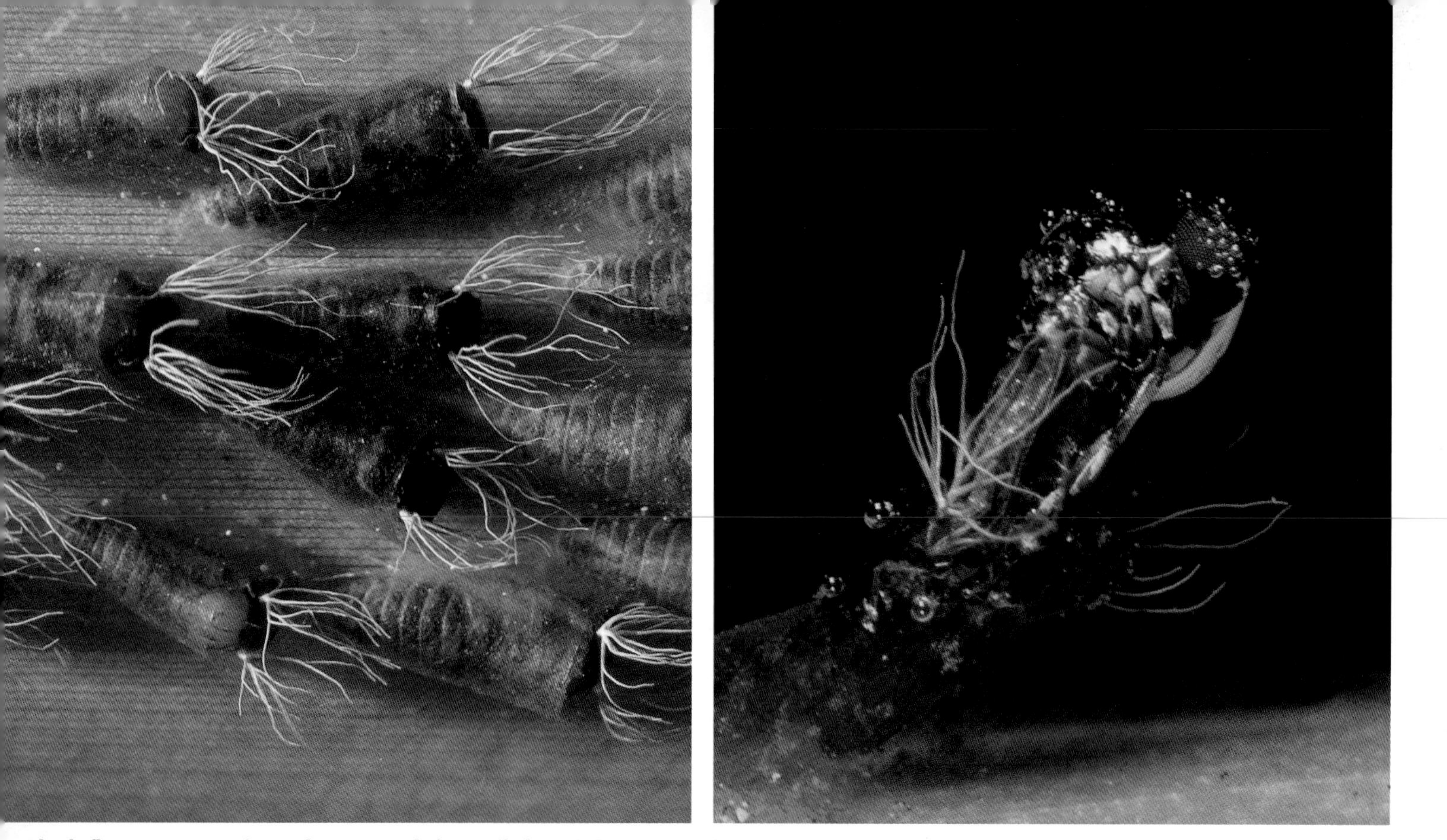

Black fly pupae remain under water (left) until the adult emerges from its pupal skin (right).

Fly larvae spend most of their time eating. Their food may be garbage, manure, algae in ponds, or other material, depending on the type of fly and where it lives. As a larva grows, it molts, or sheds its outer skin, several times. The larval stage lasts anywhere from a few days to more than a year. Then the larva changes into a pupa.

The pupae of mosquitoes still swim around, but most other pupae move little or not at all. Maggots form hard shells, or pupariums. Black fly larvae spin protective cocoons. Inside the shell or cocoon, a pupa changes into an adult fly.

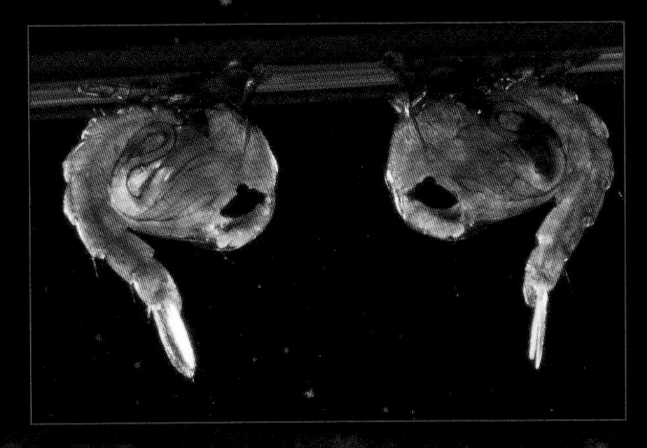

MOSQUITOES, FROM EGG TO ADULT

Inset top: **A raft of mosquito eggs floats on the water's surface.**
Inset bottom: **Mosquito pupae develop in a stagnant pond.**
An adult mosquito emerges from its pupa before flying off.

A housefly prepares to suck the sweet juice from a strawberry.

Adult flies will feed on everything from green plants to garbage, and other decaying materials. What a fly eats depends on the species. Like bees, some flies sip the nectar in flowers. Others are predators. Robber flies and grass flies, for example, catch other insects by the wings, holding them still with their powerful claws. Then they suck all the fluids from their captured prey.

Biting flies feed on blood. They pierce the skin of their prey with their sharp mouthparts and inject a substance that keeps blood flowing as they suck. With some fly types—including mosquitoes, black flies, and biting midges—only females bite. Males feed only on flower nectar. But in other species—including tsetse flies—both males and females are bloodsuckers.

Male and female flies usually mate soon after they become adults. In some species, males attract females by performing a mating flight—a sort of dance in the air. After mating, the females lay eggs and then begin the cycle again.

Not all flies live long enough to reproduce, however. Flies and their larvae are food for many other kinds of animals, including birds, spiders, and other insects.

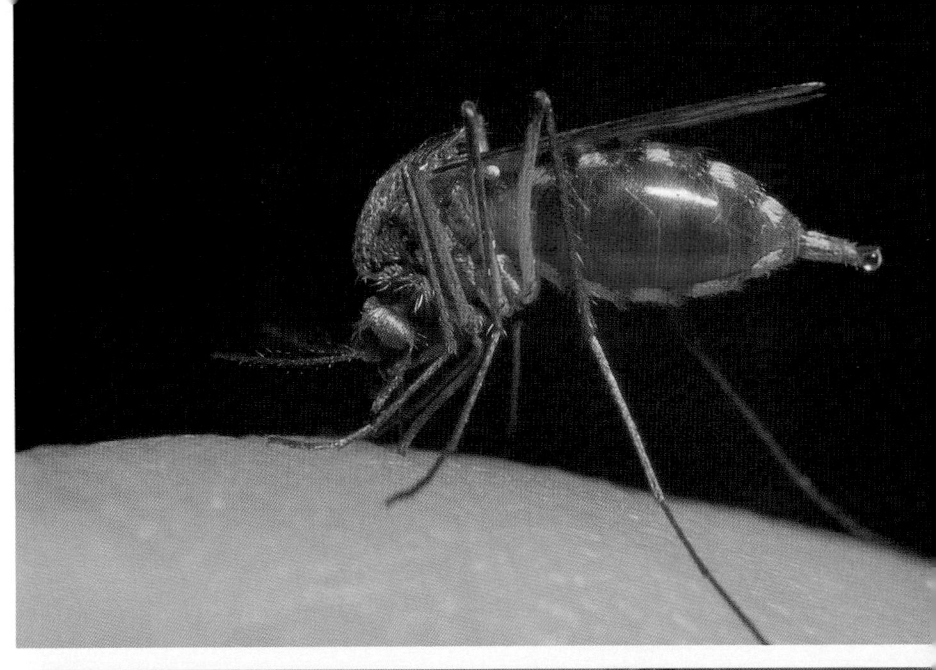

Top: **A mosquito sucks blood through its long proboscis. Its blood-filled abdomen is easy to see.** ***Bottom:*** **A grass fly consumes a black fly it has captured.**

FLIES IN DISGUISE

This hover fly looks a lot like a bee.

Some flies gleam with metallic shades of blue, green, or copper. Bluebottle and greenbottle flies, which are types of blowflies, are good examples. But as a family, flies aren't very colorful. Most are plain brown, black, gray, or dusty tan. They blend in with their surroundings, so it's hard for predators to see them. Several types of flies have another way to avoid predators. They have bright yellow or white stripes and other markings that mimic the markings of bees and wasps. Predators avoid these "flies-in-disguise" as they would avoid stinging bees.

FLIES AND PEOPLE

To people, flies are more than an annoyance—certain types of flies can actually spread serious diseases. Some flies carry disease-causing viruses, bacteria, or protozoa and transfer them to people by biting through skin. Mosquitoes may carry malaria, encephalitis, yellow fever, or other deadly diseases. Black flies in North America are not known to carry disease, but in some parts of the world they carry a tiny parasite that causes blindness. Some tsetse flies, which live in Africa, carry African sleeping sickness.

Flies that don't bite may also spread disease. For example, houseflies spread bacteria that can cause skin and digestive illness. Even the larvae of some flies can cause trouble, too. Some larvae are parasites, while others, such as fruit fly larvae, damage crops.

Above: **Some common mosquitoes, like other insects, help to pollinate flowers.** ***Left:*** **Many flies become food for other animals. Here, a jumping spider consumes a deer fly it has captured.**

For this reason, it's not surprising that most people have declared a war on flies. Insecticides are used to keep flies and their larvae out of homes and away from crops. But those chemicals can also harm the environment. Other ways to control flies include installing screens and fly traps, and releasing predators and parasites that will attack flies. Simply cleaning up food and covering garbage can control houseflies.

But not all flies are harmful. Some flies help pollinate flowers. Others attack harmful insect pests. Fly larvae are part of the natural clean-up squad that disposes of decaying plant and animal matter. And one type of fruit fly is widely used in classrooms and laboratories to study heredity. These fruit flies have helped scientists understand how traits are passed along from one generation to the next.

2

Collecting and Caring for Flies

In general, flies aren't easy to study in the wild. They're small, they move fast, and some types bite. There are, however, two types of flies that can be kept for study—fruit flies and houseflies. In order to study them and keep them correctly, you must provide the food and living conditions they need. This section will tell you where to get flies and how to keep them.

Small size and great speed make flies hard to catch in the wild.

Male fruit fly with normal wings and red eyes

FRUIT FLIES

Fruit flies are often used in laboratory studies because they reproduce and grow quickly. You can buy fruit-fly cultures from a biological supply house, such as those listed on page 46. Or, in warm weather, you can catch your own fruit flies.

A FLY TRAP

To catch fruit flies, you'll need a jar with a lid and bait. These little flies are attracted to decaying fruit, so put a banana peel, an apple core, a grapefruit rind, or something similar in the jar. Put the open jar outside, in a shady area so sunlight does not dry the fruit out. A spot near garbage cans or a compost pile is best. Lay the jar on its side, with the open end slightly downhill from the base. Check the jar in a day or so. If fruit flies are inside, quickly put on the lid.

A jar with decaying fruit makes a good fly trap.

OBSERVING FRUIT FLIES

To make the flies hold still so you can look at them, put the jar in the freezer for a few minutes. Check often—about every 20 seconds—and take the jar out as soon as you see that the flies are no longer moving. This brief exposure to cold won't hurt the flies, but don't leave the jar in the freezer for more than a minute. You can also temporarily knock the flies out with carbon dioxide gas. Biological supply companies sell kits that contain everything you need to make a carbon dioxide anesthetizer.

When the flies are "asleep," tip them out onto a sheet of white paper so you can see them better. A magnifying glass will help you get a closer look. Use a soft brush, such as a watercolor paint brush, to move them around. When the flies begin to twitch their legs and wings, they're waking up. Carefully put them back in the jar and close the lid.

You can put flies "to sleep" by placing them in a freezer for a few minutes.

KEEPING FRUIT FLIES

If you want to keep fruit flies long enough to observe their life cycle, you'll need to provide food and a place where they can reproduce. You can buy fruit-fly vials from biological supply companies. These containers have spongy plugs that allow air to circulate, and a bit of netting where flies can rest. Or you can make your own fruit-fly homes from baby-food jars. Here's how to set up a fruit-fly home:

Supplies purchased from supply companies make it easy to raise fruit flies.

What to Do:

1. Mix equal amounts of dry fruit-fly food, called instant drosophila medium, and water. Or make your own food: Mash about two tablespoons of ripe banana with 1/4 teaspoon of water. Sprinkle a few grains of yeast on top.
2. Put the food mixture in the bottom of the container. It should fill about a quarter of the jar or vial.
3. Put about ten fruit flies in the jar. Put them to sleep first, so you can transfer them easily. (See the instructions under "Observing Fruit Flies.") If you collected your own flies, check carefully to make sure they look the same, and are about the same size. They will not mate if they are not the same species.
4. Cover the container. Use a piece of fabric, secured with a rubber band, to cover a baby food jar.
5. Keep the container in a shady place. It should stay at room temperature—don't let it get too hot or too cold.

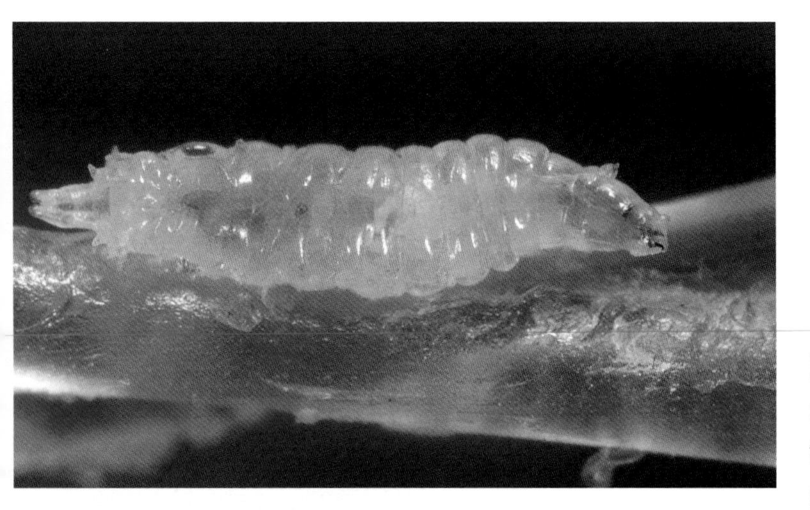

Left: **Fruit fly larva**
Below left: **Fruit fly pupa**

If the food begins to dry out, mist it with a little water. If mold starts to grow in the food, move the flies to a new container with new food. (Mold will form more quickly on homemade food.)

The flies will lay eggs in the food. When the eggs hatch, you'll see larvae moving through the mixture. Soon the larvae will form pupae on the wall of the container or the netting in the vial. The pupae look like tiny brownish cigars. About two weeks after you started the container, adult fruit flies will emerge.

If the container begins to get crowded (or the food is too dry or used up), start a new container with fresh food. Transfer at least ten fruit flies to the new container.

HOUSEFLIES

In nature, houseflies may carry diseases. You should not collect these flies from the wild. Instead, order housefly pupae from a biological supply company. They will be free of disease. Here's how to set up a simple housefly home:

What You Need:

* Plastic container, such as an empty butter dish
* Small cup or container that will fit inside the first container
* Netting
* Rubber band
* Sugar
* Powdered milk
* Wood chips (not cedar or pine) or shredded paper
* Small piece of raw meat

What to Do:

1. Mix equal amounts of sugar and powdered milk. Put some of this mixture in the bottom of the large container.
2. Put wood chips (or shredded paper) in the small cup and add water. The chips should be thoroughly soaked but still stick up out of the water.
3. Place a small piece of raw meat on the wood chips. This will provide a place for flies to lay eggs.

4. Put the cup and several dozen housefly pupae in the main container.
5. Cover the container with netting, secured with a rubber band.
6. Keep the container in a warm but shaded place. A sheltered place outdoors, or in a garage, is best. If flies escape, you won't want them in your home.

To keep and observe houseflies, order pupae from a supply company. Do not collect houseflies from the wild.

RAISING HOUSEFLIES

Part 1

After adult flies emerge from the pupae, they will mate and lay eggs. Check the meat in your container for these eggs, which will be small and white. Here's how to hatch them.

What to Do:

1. Fill the container with wood chips to within an inch of the top.
2. Mix one part powdered milk with two parts water, and pour this over the chips. The chips should be soaked but still stick out half an inch above the liquid.
3. Put the meat containing fly eggs on the chips. Cover the container with netting secured with a rubber band.

The eggs will hatch in about a week. To observe the larvae, cover the sides of the container with dark paper. When you remove the paper, you may see larvae at the sides of the container. When the larvae begin to crawl up the sides of the container to the top, they are ready to become pupae.

Part 2

Now that the larvae are ready to become pupae, here's how to collect them.

What to Do:

1. Line a large container with paper towels. Place pencils on the towels to form a base for a smaller container.
2. Transfer all of the contents of your larvae container to the small shallow container.
3. Add enough water to the small container to make the wood chips very moist, without flooding the container. Set it on the pencils in the large container.

The added moisture will cause the larvae to leave the small container. They'll collect on the paper towels and form pupae there. Put the pupae in a new fly home, and a new generation of adults will emerge.

Fly pupae

3

Investigating Flies

This chapter contains some activities that will help you learn more about fruit flies and houseflies. The first activity is done with fruit flies that you find outdoors. Others are done with flies that you collect, or with cultured flies—those that you purchase and raise.

To slow the flies down so that you can move them from one container to another, put them to "sleep" as described in Chapter 2, under "Observing Fruit Flies."

WHICH FRUITS DO FRUIT FLIES LIKE BEST?

Fruit flies are attracted to fruit, of course. But which fruits are their favorites? Make a prediction based on what you know about these little flies. Then do this activity to find out if you're right.

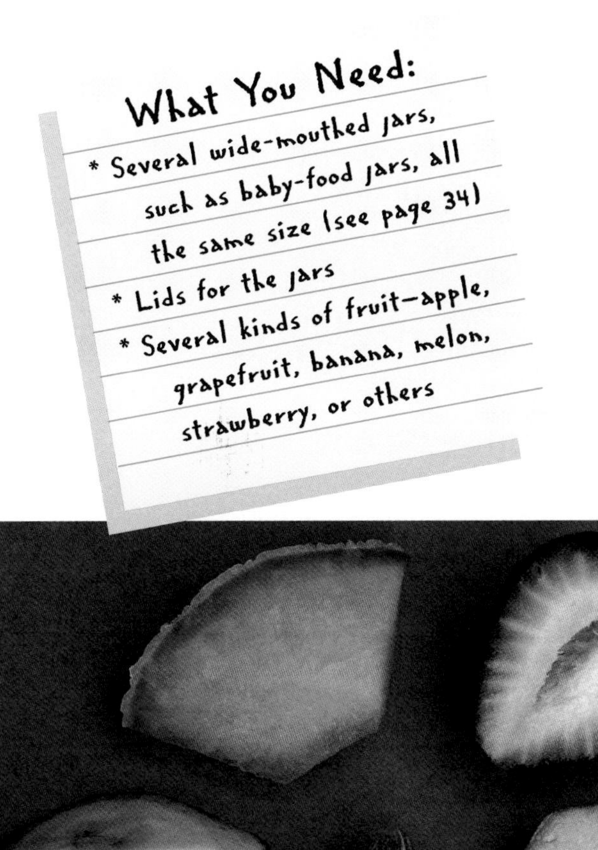

What to Do:

1. Cut the fruits into pieces, all the same size. Put the same amount of a different fruit in each jar.
2. Put the jars outside, uncovered, in the shade. Lay them on their sides, with the open ends a bit lower than the base. If possible, pick a place near a garbage can or compost pile.
3. Check the jars in a day or so. Approach quietly, and quickly put on the lids.

Results: Count the flies in each jar. Which jar has the most?

Conclusions: What do your results tell you about the fruits that fruit flies prefer? Repeat the activity, to see if you get the same results. Then try it again with different kinds of fruit, or with fruit of varying degrees of ripeness.

ARE FLIES ATTRACTED TO LIGHT?

If you've watched flies in nature, you may have an idea about whether they are drawn to light or not. Do this activity to confirm your guess.

What to Do:

1. Transfer the flies into the container, and cover the opening with plastic wrap secured by a rubber band. Use a pin to punch tiny holes in the plastic wrap, so air can get in.
2. Take the container into a dimly lighted place. There should be no strong light from windows or lighting fixtures.
3. Set the container on its side, and shine the flashlight on one end. (Use a flashlight rather than a reading lamp, which produces heat.)
4. After a few minutes, move the flashlight to the other end.

Results: Where did the flies go when you turned on the flashlight? Did they move when you moved the light?
Conclusions: Was your prediction about flies and light correct?

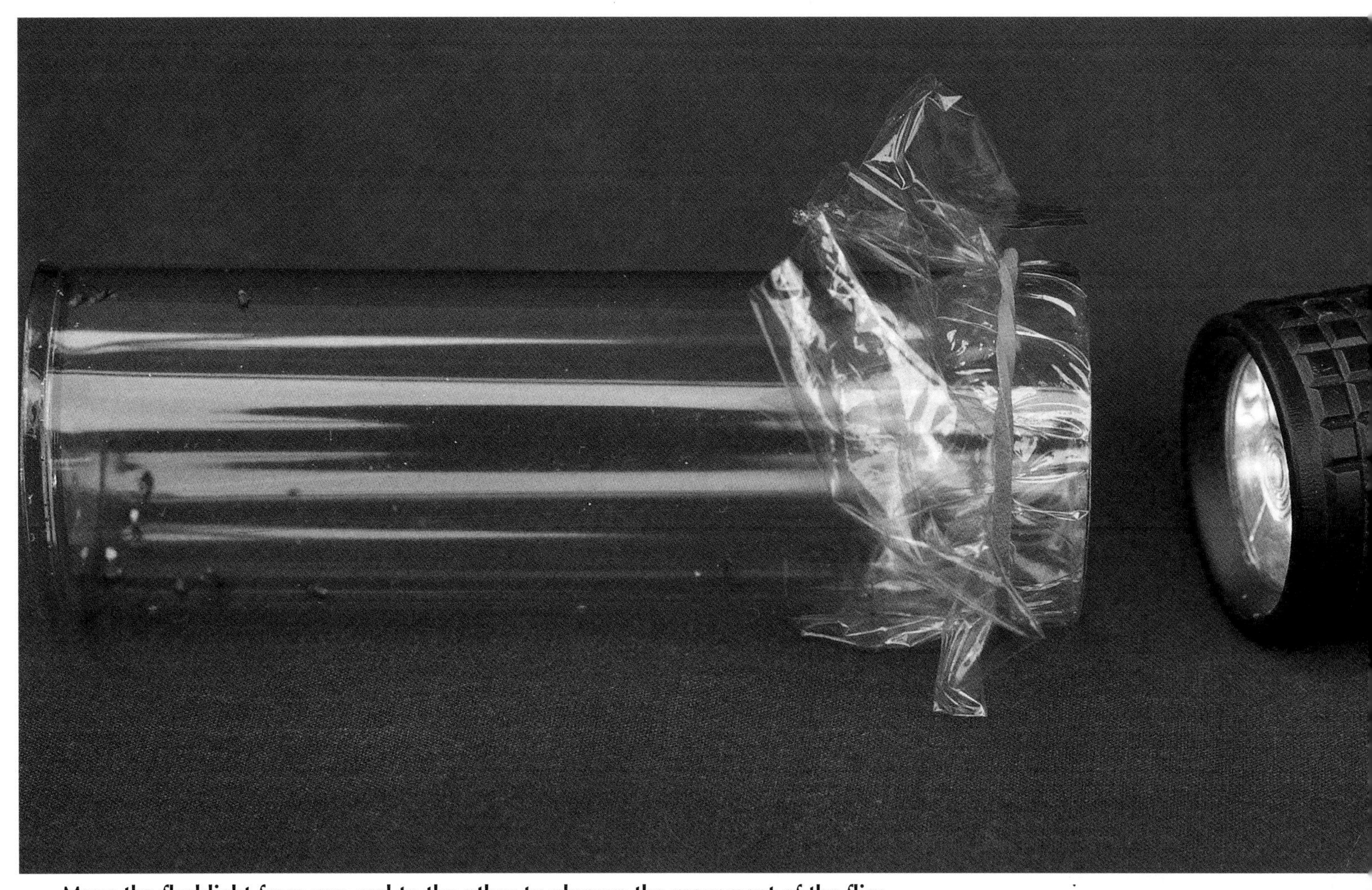

Move the flashlight from one end to the other to observe the movement of the flies.

What You Need:

* About two dozen collected or cultured fruit flies, or cultured houseflies
* Two clean vials, or other tall clear containers with tops, both the same size

DO FLIES MOVE UP, DOWN, OR ALL AROUND?

Flies are always on the go. Do you think they prefer to move down toward the ground or up into the sky? Or do they just fly around in any direction? Make a guess based on what you know about flies, and then do this activity to find out.

What to Do:

1. Put an equal number of flies in each container.

2. Watch for a while. Note where flies are located in each container.

3. Turn one vial upside down, and leave the other right side up. Note where the flies are.

4. Put both containers in a dark place. Wait 5 to 10 minutes, and then check to see where the flies are.

Results: Where did the flies go? Did turning a container upside down or placing the flies in darkness make a difference?
Conclusions: Did your flies prefer the top or bottom of their container? Repeat the experiment to see if you get the same results.

What You Need:

* Two fruit-fly cultures from a supply company. Order red-eyed females with normal wings, and white-eyed males with no wings.
* Fruit-fly vials or baby-food jars and fruit-fly food (see "Keeping Fruit Flies" in Chapter 2)
* Soft brush, such as a water-color paint brush
* Magnifying glass
* White paper

IF WINGED FLIES MATE WITH WINGLESS FLIES, WILL THE OFFSPRING HAVE WINGS?

The traits of fruit flies, like those of all living things, are determined by genes—structures inside body cells. Every fruit fly gets half its genes from each of its parents. If one parent has wings and the other does not, will the offspring have wings? Decide what you think, and then do this two-part activity. You will need fruit flies that have been specially raised to show certain traits. You can order them from a biological supply company.

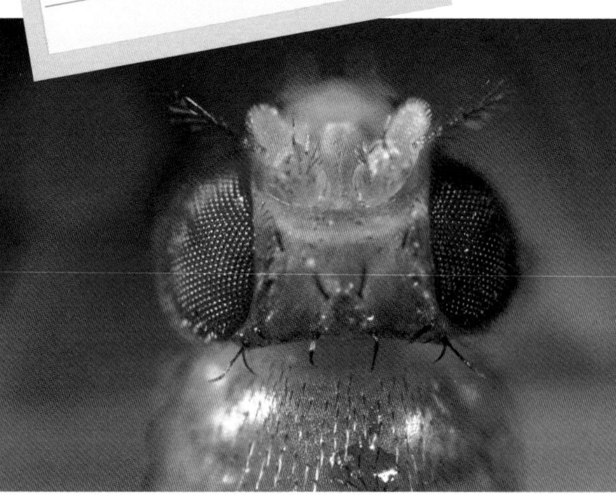

Red-eyed female fruit fly

White-eyed male fruit fly

What to Do:

Part 1

1. Put the female flies to sleep (see "Observing Fruit Flies" in Chapter 2). Carefully tip them out of their container onto a sheet of white paper. Put six of them in a new container, set up with food. Return the others to the original container.
2. Do the same with the male flies, and put six males in the new container with the six females. Label the new container. Keep it at room temperature.
3. After 7 to 10 days, the flies will have mated and the females will have laid eggs. Remove the adults.
4. In another week, a new generation of adults will begin to emerge. Put the flies to sleep, tip them out on white paper, and examine them with the magnifying glass. Count the flies, and note whether they have wings. Write down what you see. Don't return the flies to the container. Put them in a separate container for use in Part 2 of this activity, or dispose of them (you can flush them down the toilet). These specially bred flies should not be released.
5. Check the container for new flies every 2 days, for a 10-day period. Each time, repeat step 4 and record your observations.

Part 2

What do you think will happen when your flies from the new generation breed? Will their offspring—the "grandchildren" of your original flies—have normal wings? Repeat the experiment to find out. Follow steps 1 through 5 above, using the flies you raised. You will be able to tell males and females apart because the males will have white eyes, and the females will have red eyes.

White-eyed wingless male fruit fly

Results: How many flies in the first generation were wingless? How many in the second generation?
Conclusions: Do you see a pattern in the way the "wingless" trait is passed down? You can chart your results in order to observe the patterns more clearly.

Adult housefly emerging from its pupa.

MORE ACTIVITIES WITH FLIES

1. Study a housefly pupa. Remove a pupa soon after it forms, and check it every day to see how it changes. Make a journal, recording what you see. When the pupa becomes dark in color, watch closely—you may see the adult fly emerge. This usually takes five to six days at room temperature.
2. Watch a housefly eat. Place the fly in a container with some ripe fruit. Watch through a magnifying glass as it sucks up food with its long proboscis.

RESULTS AND CONCLUSIONS

Here are some possible results and conclusions for the activities on pages 35 to 43. Many factors may affect the results of these activities. If your results differ, try to think of reasons why. Repeat the activity with different conditions, and see if your results change.

What Fruits Do Fruit Flies Like Best?

Your results will vary depending on the fruits you use and how ripe they are. Ripe and overripe fruits usually draw the most flies, and bananas are a favorite with many fruit flies.

Are Flies Attracted to Light?

Because flies are drawn to light, they will probably cluster at whichever end the light shines on.

Do Flies Move Up, Down, or All Around?

Flies like to move up, against the force of gravity, so they'll probably gather at the top of the container. By placing them in clean containers, and placing the containers in darkness, you eliminate two things that can attract flies in other directions—food and light.

If Winged Flies Mate with Wingless Flies, Will the Offspring Have Wings?

Remember that the flies get genes from both parents. Thus all the flies in the first generation carry genes for "wings" and "no wings." But they all have wings because the "wings" gene is dominant. It overrules the "no wings" gene, which is recessive.

In the second generation, the results depend on which genes each offspring gets. Offspring that happen to get two "wings" genes will have wings. So will those that get one "wings" and one "no wings" gene. But those that happen to get two "no wings" genes will be wingless. If you breed enough flies and count them all, you'll probably find that the wingless flies make up one-fourth of the total in the second generation.

SOME WORDS ABOUT FLIES

Dormant Inactive.
Exoskeleton The hard outer skin of an insect. It takes the place of an internal skeleton.
Halteres Stalk-like organs, located behind the wings, that help flies balance in flight.
Larvae The young of flies.
Maggots Larvae of houseflies and certain other flies.
Proboscis Tube-like structure through which flies suck liquids.
Pupae Dormant stage during which larvae mature into adult flies.
Spiracles Holes through which flies breathe.

SOURCES FOR FLIES AND SUPPLIES

You can buy fruit flies and houseflies through the mail. Insects bought through mail-order sources such as these should not be released into the wild.

Carolina Biological Supply
2700 York Road
Burlington, NC 27215
(800) 334-5551

Connecticut Valley Biological
82 Valley Road, P.o. Box 326
Southampton, MA 01073
(800) 628-7748

FOR MORE INFORMATION

Books

Berger, Melvin. *Flies Taste With Their Feet: Weird Facts About Insects* (A Weird-But-True Book). New York, NY: Scholastic, 1997.

Goor, Ron. Nancy Goor (Contributor). *Insect Metamorphosis: From Egg to Adult.* Old Tappan, NJ: Atheneum, 1990.

Kneidel, Sally. *Creepy Crawlies and the Scientific Method: More than 100 Hands-On Science Experiments for Children.* Golden, CO: Fulcrum Publishing, 1993.

Kneidel, Sally Stenhouse. *Pet Bugs: A Kid's Guide to Catching and Keeping Touchable Insects.* New York, NY: John Wiley & Sons, 1994.

Miller, Sara Swan. *Flies: From Flower Flies to Mosquitoes* (Animals in Order). Danbury, CT: Franklin Watts, 1998.

Watts, Barrie. *Fly* (Stopwatch Series). Morristown, NJ: Silver Burdett Press, 1997.

Web Sites

A Fly Family

Meet and learn about the members of one of the main fly families, including horse flies, snipe flies, robber flies, soldier flies, bee flies, and dance flies—**www.insect-world.com/main/brachycera.html**

Flies!

Find information on many different types of flies, as well as great photographs and links to other fly sites—**www.planetpets.simplanet.com/plntfly.htm**

RobotZoo

Learn some interesting fly facts at this exciting site—**www.sgi.com/robotzoo/fly.htm**

INDEX

Note: Page numbers in italics indicate pictures.

MAY 2001
Received
Salinas Public Library